LEGENDS OF WARFARE
GROUND

M551 Sheridan

The US Army's Armored Reconnaissance / Airborne Assault Vehicle from Vietnam to Desert Storm

DAVID DOYLE

SCHIFFER MILITARY
4880 Lower Valley Road   Atglen, PA 19310

Designed by Justin Watkinson
Type set in Impact/Minion Pro/Univers LT Std

ISBN: 978-0-7643-5821-0
Printed in China

Published by Schiffer Publishing, Ltd.
4880 Lower Valley Road
Atglen, PA 19310
Phone: (610) 593-1777; Fax: (610) 593-2002
E-mail: Info@schifferbooks.com
www.schifferbooks.com

For our complete selection of fine books on this and related subjects, please visit our website at www.schifferbooks.com. You may also write for a free catalog.

Schiffer Publishing's titles are available at special discounts for bulk purchases for sales promotions or premiums. Special editions, including personalized covers, corporate imprints, and excerpts, can be created in large quantities for special needs. For more information, contact the publisher.

We are always looking for people to write books on new and related subjects. If you have an idea for a book, please contact us at proposals@schifferbooks.com.

Acknowledgments

While my name is the one that is on the cover of this book, the truth is that this book could not have been created by any one person. The Lord has blessed me with a number of friends who have graciously and enthusiastically given of their time to make this book a reality. This includes Tom Kailbourn, Chris Hughes, Mike Koening, and Scott Taylor. I am especially thankful to have received the help of several veterans, including Doug Kibbey, Chris Harlow, and William Powis. Additionally, the staff and volunteers at the Patton Museum and the staff at the National Archives and TACOM provided invaluable assistance. Most importantly, I am blessed to have the help and support of my wife, Denise, for which I am eternally grateful.

All photos of preserved vehicles were taken by the author, unless otherwise noted. Front cover photo by Chris Hughes.

Contents

Introduction

The M551 Sheridan, named after Civil War–era Union general Philip Sheridan, was a controversial vehicle almost from the time production began on July 29, 1966, at the Cleveland Tank Plant. To some, the Sheridan is seen as an extremely complicated, and expensive, weapons system that was minimally effective. Others see the Sheridan as a machine that despite some teething problems and expense delivered to troops capabilities that not only were unequaled but have been lacking in the US military since the M551's retirement from combat units in 1996.

The genesis of the Sheridan stretched all the way back to 1945. In November of that year, the 1945 War Department Equipment Board (popularly known as the Stillwell Board) began laying out criteria for the post–World War II development of Army materiel of all types, including tanks. Using as a baseline the then-current US light tank, the M24 Chaffee, the Equipment Board laid out the design parameters of future light tanks as such: "The light tank is for reconnaissance and security. (2) The present light tank should be further developed to obtain the maximum cross-country mobility and maneuverability and all[-]around armor protection against small arms and shell fragments, with frontal armor protection against light antitank guns. It should not exceed twenty-five tons in weight. The tank gun should be of approximately three-inch caliber, capable of penetrating five inches of homogenous armor at 30-degree obliquity at 1,000 yards using special ammunition."

Further, the words of the War Department Equipment Board were not alone in their conclusions; the Army Ground Forces Equipment Review Board (Cook Board), in June 1945, had much the same findings, as did the report of the Armored Equipment Board (Robinett Board), whose 1945 work at the Armor Center at Fort Knox generated a voluminous report.

In July 1946, designers began to lay out a vehicle that would achieve the lofty goals of the War Department Equipment Board, working their way through a series of designs that would result in the M41. The M41, named the Walker Bulldog, was a notable improvement over the M24 and was well liked by tank crews, but it fell short of the Stillwell Board goals. The particular shortcomings were a lack of range and being overweight. Thus, the M41 was considered only an interim vehicle by Army planners.

In May 1952, the characteristics of a successor to the M41 were laid out. Counter to the usual upward creep of combat vehicle weight, the new vehicle's upper weight limit of 20 tons was revised downward to 18 tons. After circulating the criteria among potential bidders, proposals were received from Detroit Arsenal, the Cadillac Motor Car Division of General Motors, and Aircraft Armaments Incorporated (AAI) by July 1953.

The AAI design was considered the winner, and production of a pilot model, designated the T92 76 mm gun tank, was initiated. The AAI engineers had designed a very compact vehicle, reasoning that the lesser volume would require less heavy armor to adequately protect, as well as presenting a smaller target. The first pilot model arrived at Aberdeen Proving Ground for testing in November 1956. The results of the pilot model testing went well, and it was anticipated that after further development the type would enter production in 1962. However, that would not be the case.

Rushed into production amid concerns that the Korean War would escalate, the M41 Walk Bulldog, seen here on the production line at the Cleveland Tank Plant, represented a significant step forward from the World War II–era M24 Chaffee. However, the M41 fell short of ideals set by the Army in 1945. *Patton Museum*

Although extensive testing of the T92 light tank was being undertaken, the revelation that the Soviets were developing a new amphibious light tank, the PT-76, caused Congress to demand an American counterpart. Thus, plans for the T92 were scrapped, and a new amphibious vehicle was designed, the XM551, seen here in a late development sketch. *Patton Museum*

GENERAL SHERIDAN XM551

CHAPTER 1
Development

In 1957, a congressional committee looking at military matters noted that the Soviet Union was also outfitting its force with a new light tank—but theirs was amphibious. The US army was then challenged: Why was its new light tank that was being proposed for production not also amphibious? At this point the compact design of the T92 began to work against its future. In order for a vessel to float, it must weigh less than the volume of water that its volume displaces. The T92 design, with its small volume and relatively high mass, not only would not float, it could not reasonably be made to float. Accordingly, in June 1958, further development of the T92 was canceled. The engineers returned to their drawing boards, now adding amphibious capabilities to the long list of criteria laid out by the Stillwell Board.

The new vehicle was created at a time that the Army was undergoing changes in its classification systems. No longer were experimental vehicles given a "T" designation; rather, they were assigned an "X" prefix to their proposed standard model designation. Further, no longer were tanks being classified as light, medium, and heavy—instead, the Army had adopted the Main Battle Tank concept, wherein a single tank would be used.

Accordingly, the replacement for the T92 Light Tank was the XM551 Armored Reconnaissance / Airborne Assault Vehicle. It looked like a tank, GIs called it a tank, but officially it was not a tank. As the name indicated, the XM551 was designed to be a light vehicle with both amphibious and airborne assault abilities.

In order to achieve the light weight, the hull of the vehicle was made of welded 7039 aluminum-alloy armor plate. While this indeed helped achieve the desired weight, when the Sheridan first appeared in a combat environment, the aluminum hull was known to be extremely vulnerable to antitank mines, as was the M113 armored personnel carrier, which also featured a welded aluminum hull. Because of prior experience with the M113, the Ordnance Tank-Automotive Center quickly developed a hull armor kit to be fitted to those vehicles that were sent to Vietnam.

While the hull of the Sheridan was aluminum, the turrets were made from steel armor. That turret housed three of the four men who formed the vehicle crew: the commander, gunner, and loader. The driver was located in the front center of the hull. The hull, in addition to being made of aluminum enrobed in a layer of high-density foam, was itself encased in the aluminum skin, to further improve flotation. Powering the vehicle was a six-cylinder Detroit Diesel engine.

While the vehicle was developed by the Cadillac Division of General Motors, by the time series production began on July 29, 1966, responsibility for the project had been transferred to the Allison Division of General Motors. Allison produced 1,562 of the vehicles in the Cleveland Tank Plant before production ceased in 1970.

Cadillac delivered to the Army the first of twelve pilot vehicles for its proposed amphibious, airdroppable armored reconnaissance / airborne assault vehicle (AR/AAV), the XM551, in June 1962. While Aircraft Armaments Incorporated and Allis-Chalmers also submitted a proposal for an AR/AAV, it was Cadillac's version that the Army accepted. The pilot vehicles were designated the XM551 Sheridan. The first pilot, shown here, had a boxy hull, bogie wheels with embossed "spokes," a trim vane or surfboard on the glacis, with a window in it to assist the driver, and a turret with a sharp edge partway up the sides and rear. *TACOM LCMC History Office*

The first three XM551s comprised the first-generation pilots. These had the box-shaped rear hulls, bogie wheels with embossed spokes, and continuous-band tracks. At the heart of the XM551's offensive system was the XM81 gun launcher, capable of firing a conventional 152 mm round or an XM13 Shillelagh guided missile. The platform on the upper part of the gun shield, or mantlet, was for mounting the optical transmitter of the missile sight / missile-tracking system, not yet available when the photo was taken. *Patton Museum*

For operating on water, the first- and second-generation XM551 pilots were equipped with a jet water-propulsion system. Visible on the left rear of the lower hull of the first pilot XM551 is the left water-jet propulsion nozzle, with a similar one faintly visible on the right side. Between them is the water intake for the propulsion system. *Patton Museum*

XM551 pilot number 3 was employed in the first firing tests of the complete XM551 Sheridan / XM81 Shillelagh system at White Sands, New Mexico. The missile fired by the XM81 Shillelagh system was the XM13 Shillelagh, a folding-fin-stabilized guided missile, controlled by the gunner through an infrared tracking-and-command system. In the background is a battery of tracking equipment, including motion-picture cameras on a stand above the turret. *Rock Island Arsenal Museum*

One of the three first-generation pilot XM551s is mounted on a pallet and poised for dropping by a crawler crane by the Air-Test Branch at the Yuma Proving Ground, Arizona, around the early 1960s. This test was to evaluate the adequacy of the structure of the vehicle in controlled-impact tests, simulating the effects of a real airdrop landing in various conditions. *National Archives*

Pilot number 4 of the XM551 Sheridan was the first of three second-generation XM551s, characterized by, among other things, new, 24½-inch-diameter bogie wheels; new, single-pin, link-type tracks; a windowless trim vane; and a greatly revamped hull with a sloped rear. The fourth pilot XM551, registration number 12E289, is seen here during fording tests, with trim vane and flotation screens erected. Flotation bags were behind the trim vane and, partially visible here, on the rear of the hull. *US Army, courtesy of 1st Sgt. Vic Pitts via Doug Kibbey*

The windowless trim vane of XM551 pilot number 5 is displayed in the lowered position. The headlight assemblies are above the trim vane. The new, single-pin tracks were formed of separate links. The bulge on the upper half of the left side of the gun shield contained an opening for a spotting rifle (not installed here), which had the same trajectory as the 152 mm HEAT-MP round at out to 1,300 yards. A coaxial 7.62 mm machine gun was in the shadow under the spotting rifle. The box on the top of the gun shield contained the optical transmitter, part of the missile sight / missile-tracking system. *TACOM LCMC History Office*

XM551 Sheridan registration number 12Z292, seen in a November 1963 photo, was the first of the third-generation pilots. Changes since the second generation included a significantly redesigned hull front and trim vane; a new, rotating driver's hatch with three periscopes; flotation covers on the bogie wheels; elimination of the jet water-propulsion system in favor of track-powered propulsion in the water; and a recessed activating handle for the fire extinguisher near the front of the left sponson. *Rock Island Arsenal Museum*

The seventh pilot XM551, registration number 12Z292, is being test driven over rough terrain. The new, rotating driver's hatch is closed. A manually operated Browning .50-caliber M2 HB machine gun is mounted on the cupola. *National Archives*

In October 1963, the seventh pilot XM551 is being tested for its ability to negotiate vertical obstacles. The maximum vertical wall was 18 inches for the first-generation XM551s, while this increased to 33 inches for production M551 Sheridans. The storage of machine gun ammunition boxes on the sides of the turret would remain a feature on Sheridans into the future. *Rock Island Arsenal Museum*

The seventh pilot XM551 maneuvers over a test track during evaluations. The flotation covers have been removed from the bogie wheels. Stenciled in white on the sponson is "U.S. ARMY / 12Z292 / XM551 / PILOT NO. 7." *TACOM LCMC History Office*

Pilot number 8 of the XM551 varied from pilot 7 in that the two vehicles had different levels of armored protection. Seen here with its flotation equipment deployed and watertight seals installed on the gun launcher, pilot 8 bore registration number 13C503. The trim vane on this vehicle had a window in the center. *US Army, courtesy of 1st Sgt. Vic Pitts via Doug Kibbey*

An experimental water barrier, painted in horizontal stripes, has been installed on the eighth pilot XM551, which is about to test the rig. A horizontal white line has been painted on the vehicle however many inches above or below the established waterline the vehicle is situated when fully afloat. *US Army, courtesy of 1st Sgt. Vic Pitts via Doug Kibbey*

The ninth pilot XM551, registration number 12FD25, is viewed from above with the flotation gear erected. The trim vane was a single-piece unit with a central window and two braces to the rear. Note the extension trunk on the engine exhaust at the rear of the engine deck. Part of the driver's rotating hatch is visible, as are details of the turret, including the commander's cupola with its ten vision blocks and split hatch doors; the gunner's periscope to the front of the cupola; the loader's hatch and separate rotating periscope; and, on the side of the turret below the loader's hatch, the ventilator hood. *US Army, courtesy of 1st Sgt. Vic Pitts via Doug Kibbey*

Pilot number 10 of the XM551 Sheridan, registration number 12FD26, test-fires its gun launcher. This vehicle, as well as pilots number 9 and 11, were delivered to the Army in 1964, with engineering and service tests commencing in October of that same year. *TACOM LCMC History Office*

The final pilot XM551, registration number 12FD28, was delivered during February 1965, and it embodied the planned configuration of the M551 Sheridan. This vehicle included a revamped flotation system, with the flotation barrier stowed under the curved covers at the tops of the sponsons and across the rear of the hull, and the single-panel trim vane replaced by a new, bifolding version. *Patton Museum*

A right-rear view of an XM551 provides an excellent idea of the design of the flotation barrier when erected. The covers for the stowed barrier are open, and the straps with snaps that secure the covers when closed are flapping loose. At the front of the vehicle, the upper part of the trim vane and its window are visible. At intervals, support posts hold up the barrier. Visible above the top of the rear of the barrier are the two rear bilge-pump outlets. *Patton Museum*

XM551 pilot number 12 is fording a body of water. The two curved rear outlets for the bilge pumps are visible at the rear of the flotation barrier, and there also was a front bilge-pump outlet, which is seen here about 1 foot to the front of the forward whip antenna. *US Army, courtesy of 1st Sgt. Vic Pitts via Doug Kibbey*

The twelfth pilot XM551 is loaded on a flatcar at Fort Lee, Virginia, on March 7, 1966. The vehicle was undergoing movement-adaptability tests at the time. *National Archives*

During a test of cross-country performance, the twelfth pilot XM551 splashes across Otter Creek at Fort Knox, Kentucky, on May 6, 1965. The spotting rifle had been discontinued, and its port in the gun shield had been plugged. *National Archives*

The twelfth pilot XM551 was subjected to airdrop tests at Fort Knox, Kentucky, in 1965. The vehicle is seen here in its first such test after being dropped by a Lockheed C-130 Hercules. Three parachutes have deployed. What is visible here is the bottom of the pallet that held the vehicle. *US Army, courtesy of 1st Sgt. Vic Pitts via Doug Kibbey*

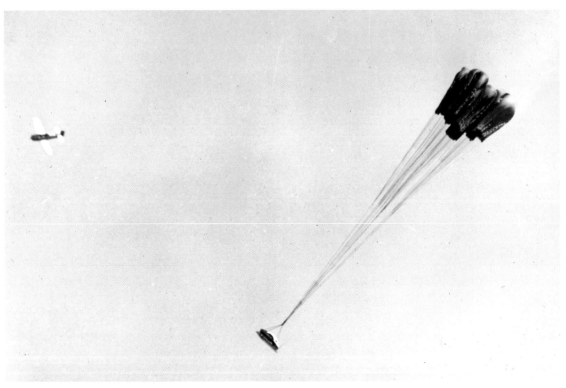

The XM551 is seen from the right side on its pallet as the parachutes deploy during a drop test in 1965. A chase plane is visible to the upper left. *US Army, courtesy of 1st Sgt. Vic Pitts via Doug Kibbey*

At a desert missile range in October 1967, XM551 pilot 12 has just fired a Shillelagh missile. After clearing the muzzle of the gun launcher, the folding fins on the missile have deployed. By now, the turret had been equipped with four smoke-grenade projectors on each side, with guards. *US Army Engineer School History Office*

XM551 number 12 is illuminated in the flash of a Shillelagh missile leaving the gun launcher during a nighttime firing test. During firing, the gunner held the crosshairs of his sight on the target. After flying the first 800 yards in unguided mode, the missile went into guided mode, in which the gunner controlled the final leg of the missile's flight by an infrared link. *Patton Museum*

The track of a Shillelagh missile is evident in this time-lapse photo taken at night as it races downrange from the twelfth pilot XM551, out of view to the left, to its target on the rise in the background. *Patton Museum*

Launch Vehicle

Target Vehicle

To facilitate gunner proficiency with the Shillelagh missile, a system of two M551s with training aids was developed. Photographed during testing, on the left is a launch vehicle, registration number 13C602, while on the right is a target vehicle, registration number 13C520. The launch vehicle was equipped with a visual-effects simulator for the gunner's telescopic sight and a box atop the turret containing the instructor's control indicator. The target vehicle had a tower atop the turret, with an infrared transmitter on top. The launch and target vehicles ultimately were designated the XM41 launcher and the XM42 target, with the combined vehicles being designated the Conduct of Fire Trainer XM35. *US Army, courtesy of 1st Sgt. Vic Pitts via Doug Kibbey*

A Shillelagh missile has just been fired from an M551 Sheridan and is visible as it races downrange to the left, during testing by the US Army Test and Evaluation Command's Armor-Engineer Board at Fort Knox, Kentucky, in October 1967. There is a white-stenciled "M551" on the sponson. *National Archives*

Accessory equipment displayed next to M551 registration number 13E962 includes such items as electronic devices and spare machine gun barrels (*upper center*); combat vehicle crewman (CVC) helmets, flashlight, signal flags, and periscope (*upper right*); pioneer tools (*lower right*); and hand tools (*lower left*). *US Army, courtesy of 1st Sgt. Vic Pitts via Doug Kibbey*

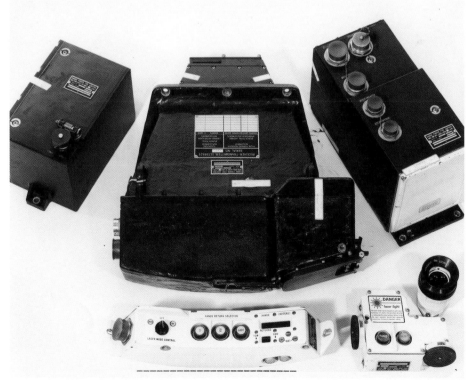

The armored reconnaissance / airborne assault vehicle (AR/AAV) M551A1 was an improved version of the Sheridan equipped with the laser rangefinder (LRF) AN/VVG-1 and the tank thermal sight (TTS) AN/VSG-2B. Elements of the laser rangefinder system included (*top row, left to right*) the battery power supply, receiver-transmitter, and laser power supply; and, *bottom*, the commander's laser-ranging control panel, the receiver/transmitter control, and eyepiece. *US Army, courtesy of Col. John Berres via Doug Kibbey*

M551 Sheridan registration number 13C503 is viewed from the right rear with the turret traversed aft at Rock Island Arsenal, Illinois, on March 10, 1967. This vehicle was used as a testbed at the arsenal for mounting several conventional artillery pieces in the turret, which became a consideration in early 1967 because of problems with the combustible-case ammunition of the 152 mm gun launcher M81. Installed here is a 105 mm Howitzer XM103E7. *National Archives*

A 76 mm Cannon M32 has been mounted in the turret of M551 registration number 13C503, as photographed at Rock Island Arsenal, Illinois, on March 24, 1967. Although the 105 mm howitzer and 76 mm cannon proved satisfactory in the M551, neither weapon was standardized in the Sheridan. In the background is the fuselage of a Piasecki H-21 Workhorse helicopter. *National Archives*

CHAPTER 2
Armament

The principal armament of the Sheridan was the 152 mm gun/launcher, which could fire either conventional ammunition or the Ford-produced Shillelagh antitank missile. This weapon system was designed primarily as a missile launcher, with the gun role being secondary. Ironically, but not necessarily surprisingly, it was the gun aspect that proved both most troublesome and most widely used.

Most of the gun problems were actually ammunition problems. Rather than using a conventional metal cartridge case, the Sheridan's ammunition (which it also shared with the M60A2 "Starship") had combustible cartridge cases. In theory, when the round was fired, the cartridge case was consumed by burning, with the ash expelled from the muzzle—thus solving the problem of handling and storing the used cartridge cases.

The nature of these cases was such that they had to be protected from moisture—even humidity—in order for the cases to stay together and completely burn. Clearly, in Vietnam humidity would be a major problem. In an effort to help with this, a removable bag system was developed, with the loader stripping the bag off just as the round was chambered. The relatively fragile nature of the rounds, added to the vulnerability of the vehicle to land mines, meant that striking a mine meant there was a real risk of ammunition detonation, leading many Sheridan crewmen to ride outside the vehicle.

When the Sheridan began to be considered for deployment to Vietnam, where there was rarely tank-to-tank combat, a significant shortcoming in the weapon system was the lack of alternative ammunition, such as antipersonnel ammunition. To answer this, the M625 flechette antipersonnel round was rapidly developed.

In order to improve accuracy, in 1971 Hughes Aircraft was contracted to create the AN/VVG-1 laser rangefinder for use on the Sheridan. Vehicles modified through the installation of this system were reclassified as M551A1. One further improvement was made in the combat capabilities of the Sheridan, this just prior to Desert Storm. This modification entailed the installation of the AN/VSG-2B Tank Thermal Sight as used in the M60A3. Sheridans so equipped were designated M551A1(TTS) and were the most-accurate versions of the vehicle fielded.

The M551 Sheridan was rolled out for the public in dramatic style at the Cleveland Tank Plant on June 29, 1966, crashing through a curtain to dramatic effect. The Allison Division of General Motors assembled the M551s at the Cleveland Tank Plant. For the rollout, some 2,000 employees of the Allison Division watched the vehicle perform a number of maneuvers on the test track. *Patton Museum*

A newly completed M551 Sheridan, registration number 13F177, with all exterior equipment in place, is parked on the factory floor of the Cleveland Tank Plant. A xenon searchlight is mounted over the gun shield, and the commander's .50-caliber machine gun is equipped with a flash suppressor on the muzzle and an AN/PVS-2 night scope above the receiver. In the left background is the hull of an MBT-70, another vehicle designed to use the Shillelagh missile. *Patton Museum*

An early-production M551 Sheridan, registration number 13C523, was photographed while undergoing testing by the US Army Test and Evaluation Command's Armor-Engineer Board at Fort Knox, Kentucky, in October 1967. The 152 mm gun launcher M81 is fitted with a bore evacuator over the barrel, a feature that later was discontinued. *National Archives*

When it was found that the bore evacuator in combination with the gun launcher's open-breech scavenger system tended to blow burning residue from the combustible-case ammunition back into the turret interior after firing, a new, closed-breech scavenger system (CBSS) was installed in M551s. This system used compressed air to blow any smoldering residue out of the gun launcher before the breech was opened after firing. At some point after the introduction of the CBSS, the bore evacuators were removed from the barrels, since they now were redundant. This M551 Sheridan, registration number 12C63268, exhibits the barrel without the bore evacuator, and there is a reinforcing cuff over the muzzle. *Patton Museum*

The gun launcher barrel with the bore evacuator removed is present on M551 registration number 13F177. A xenon searchlight is installed, and the commander's .50-caliber machine gun has a rudimentary shield, with panels on both sides that include folding flaps at the top. Both flaps are folded down in this photo. *Rock Island Arsenal Museum*

M551 Sheridan registration number 13F152 poses without the customary machine gun ammunition boxes stored on the side of the turret. The gun launcher has the bore evacuator. Note the position of the ventilator hood on the right front of the turret roof. Covers are installed over the muzzles of the smoke-grenade projectors. *TACOM LCMC History Office*

This M551 serving as a test vehicle has been modified with several features, including a shield above the commander's cupola that is similar to the ACAV shield introduced in the Vietnam War, except with a cutout on the top of the gun slot to allow for the line of sight of a night scope, and an inner shield to protect the commander's sides; a guard fabricated from rods for the xenon searchlight; and fairly substantial guards for the smoke-grenade launchers. *Patton Museum*

The experimental armored shield for the .50-caliber machine gun is visible from the right front on M551 registration number 13F177. Above the barrel of the gun launcher, the door of the optical transmitter is open. The large aperture with the guard on the front of the gun shield is the objective of the gunner's telescopic sight, which worked in conjunction with the optical tracker to monitor the missile upon firing, determining how far it has deviated from the gunner's line of sight and sending that data to the missile guidance and control system. *Patton Museum*

M551 Sheridan registration number 13C512 is driving over the hill gravel course during a dry-road test at Yuma Proving Ground, Arizona, on November 12, 1968. Markings on the vehicle signify that the vehicle was being tested by the US Army Test and Evaluation Command, Yuma Proving Ground.
National Archives

The M551 Sheridan had a long history with the US Army, serving exclusively with that branch for over thirty-five years. The National Armor and Cavalry Heritage Foundation maintains this example.
Mike Koening

Another view of the NACHF vehicle next to one of its Vietnam War stablemates: an M113 armored personal carrier. The M551 was named after Maj. Gen. Philip H. Sheridan, the famous commander of the Union army cavalry. *Mike Koening*

The object along the edge of the rear hull is the main exhaust-gas outlet. Below that are the attachment points for the pioneer tools, missing in this photograph. A grab handle is seen at the left, which is one of two used for the removal of the access panels that the tool brackets are mounted to. *Author*

The right-hand exhaust grate is seen here, with additional attachment points visible for the pioneer tools visible at the top center. Various footman loops for securing the tools with straps are also present. The access panels to the left are for the engine battery. *Author*

Hot air for the engine is vented outward through two large grates on the rear of the M551. This is the left-hand grate. Various locking pins and chains for retaining the access doors are visible here. *Author*

A portion of the flotation barrier screen is stowed along the back of the vehicle for use in water fording, and it is secured with seven canvas straps. One of two slotted bilge outlets can be seen at the top-right center of the photo. *Author*

An enclosed commander's cupola was provided that allowed the supplemental .50-caliber machine gun to be fired from relative safety. This was absent from early models of the M551. *Author*

A view of the commander's cupola from the lower left. Just in front, on the slopping surface of the turret roof, is the turret ventilator. Also seen are the two masts for the AN/VRC-12 radio set used in early-model vehicles. *Author*

As seen from the front of the vehicle, the opening to the left of the main gun is the gunner's telescopic sight aperture. The long shroud helps shield the sight from glare, as well as precipitation. *Author*

Due to the limited internal space of the M551, much of the additional gear carried was stowed on the exterior of the vehicle. Brackets and footman loops were provided for stowing ammunition cans on both sides of the turret, although most were designated for the left side. *Author*

The rear of the commander's cupola contained a door that could be folded down for egress. It was secured by a small latch and pin (*at center left*). *Author*

The .50-caliber machine gun as seen from the commander's position. The large bracket just behind the front shield would hold a tray that contained a .50-caliber ammunition can so that the belts could be fed directly into the gun. *Author*

The right side of the cupola assembly. The halves of the opened commander's hatch completed the armor protection. Note the hatch springs at the base of the hatch. The top portion of the shield assembly was removable. *Author*

This portion of the shield was held in place by a steel post and brackets. Details of the bolted mount are visible below. Pulling down on the release lever seen at lower right closes or opens the commander's hatch. *Author*

Starting in 1972, the M551 was equipped with a laser rangefinder. This is the small boxlike structure below the front of the cupola and was known as LRF ANNVG-1. Vehicles so equipped were designated M551A1. *Doug Kibbey*

As mounted in the commander's cupola, the LRF determined the distance to target in meters. This improved the first-hit capabilities of conventional ammunition. When installed, the LRF took the place of the front vision block of the cupola. *Doug Kibbey*

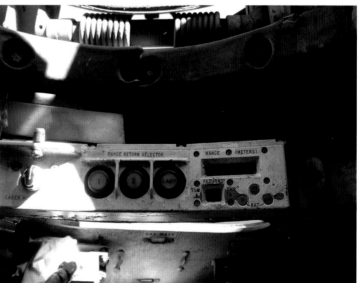

The controls and readout for the LRF are located at the commander's station to the left of the cupola, just below the rim. The system determined distance to target by projecting a laser beam and measuring the time it took for the beam to return to the vehicle. *Doug Kibbey*

In 1977, the M551 was modernized within the parameters of the Product Improvement Program. This introduced many improvements to the exterior and the power plant. This vehicle, "Deathstalker," is one such vehicle. It was in the collection of the Military Vehicle Technology Foundation for many years and is seen here being run at their facility. *Chris Hughes*

The MVTF Sheridan is marked as a vehicle from the 82nd Airborne Division as it would have appeared in the early 1990s. The 82nd was the last unit to deploy the M551 by that point, since it was the only US Army armored vehicle available that could be airdropped. *Chris Hughes*

The MVTF M551 is put though its paces in the courtyard of their facility. Copious amounts of soot-laden air are emitted from the two large grates on the engine deck. The main exhaust-gas outlet was also a source for this grimy gas, and it frequently fouled the pioneer tools. *Chris Hughes*

This three-quarter view provides a good perspective of the flotation barrier screen stowage along the edges of the tank's lower hull. This ran along three sides of the hull and contained a screen used in conjunction with a front-mounted surfboard that could all be deployed when the tank was crossing water obstacles. *Chris Hughes*

Details of the left side of the turret of the MVTF's Sheridan are displayed, including the ballistic armor surrounding the commander's cupola. Also in view are the AN/VSS3 searchlight, mounted over the front of the turret; the turret ventilator to the rear of the searchlight; the left-hand smoke-grenade launchers; and stored ammunition boxes and five-gallon liquid container. In a round recess on the side of the sponson below the left headlight assembly is a red handle for actuating the vehicle's fire-extinguisher system. *Chris Hughes*

An overall view of the left-hand side of the turret with its many .50-caliber ammunition cans in place. There are a total of six here, with a single box of 7.62 mm ammunition located at the extreme left. The rear of the vehicle is to the right. *Chris Hughes*

The engine deck with the warm-air-exhaust screens to the right and the air intake screens to the left, just under the turret. The main engine exhaust pipe is to the far right. *Chris Hughes*

With the warm-air-exhaust doors open, the armored slats became apparent. Some details of the General Motors 6V53T, six-cylinder, turbocharged Diesel 300-horsepower engine are visible here, such as the air inlet system and the fuel filters at left center. *Chris Hughes*

The exhaust doors are secured with a metal latch pin, seen located between them. Although all the various engine doors are covered with anti-debris screens, each contains armored slating to protect the vital engine components from metal fragments in the combat environment. *Chris Hughes*

This closer perspective reveals the turbocharger air inlet hose (*lower left*), the coolant surge tank (*circular object left of center*), and engine oil filler caps (*either side of divider, above center*). There was both a primary and secondary fuel filter—they are linked together and seen at the lower right of the right-hand opening. *Chris Hughes*

In order to open the large warm-air-exhaust doors, the turret must be rotated 90 degrees either to the left or right. Although the hull of the M551 was fabricated from aluminum, the turret was of steel construction. *Chris Hughes*

The AN/VSS-3 searchlight as viewed from the left side. It could project either white for general illumination or infrared light for night-fighting operations. The rear cover could be removed to service the light. The stencil on top reminds the crew that detaching the heavy assembly is a two-man job. Handles are located on either side for this task. *Chris Hughes*

A total of eight XM167 grenade launchers were mounted on the turret, with four on each side. A master switch installed in the commander's station controlled them. He could elect to fire them in various combinations, or all at once. *Chris Hughes*

This view from the right side of the gun mantlet shows the AN/VSS-3 searchlight in the background and the optical transmitter box at center. This device transmitted infrared signals to the Shillelagh missiles in order to guide them to their target. *Chris Hughes*

The commander's cupola as seen from the right side of the turret. The front of the turret is to the right side of the photo. The .50-caliber machine gun is not installed here. The sun, at lower center, highlights the bright glass of the commander's vision blocks. *Chris Hughes*

This left-hand photo provides a good overview of the layout of the armor panels on the commander's cupola. The top-front panel as well as the two smaller panels on the left and right could be folded down. *Chris Hughes*

This view from below shows the massive triangular support mount for the .50-caliber machine gun within the cupola's armored assembly. A metal plate located in the void of the base protects the commander from shrapnel. *Chris Hughes*

Looking down into the commander's hatch, the controls are visible on the interior of the hatch halves. The rounded levers are the hatch cover release levers, while the straight levers on the left half are the hatch cover locking levers. *Chris Hughes*

Looking back at the commander's position, the folding door is seen. Just visible in the rear corners are two additional footman loops that can be used to stow another two .50-caliber ammunition boxes on either side. *Chris Hughes*

The loader's hatch on the left side of the turret. The loader's M37 periscope is located forward of that, to the right of the photo. Just to the left of the hatch is the base of the AS-1729NRC antenna. *Chris Hughes*

The right shield assembly is seen here. It could be folded down by first removing the shield bar lock (a simple cotter pin) on the shield bar and then lifting the bar straight out. A small strip of steel protected the gap between both the right and left shield assemblies and the main armored cupola unit. *Chris Hughes*

The main turret ventilator is to the right of center, with the loader's M37 periscope to its left and the armored conduit for the searchlight power cable to its right. The .50-caliber ammunition tray and box can be seen at lower right. *Chris Hughes*

Inside the turret of the Sheridan, the 155 mm gun launcher is viewed from its left side with the breech removed. To the right of the photo is the recoil guard, which includes expanded steel-mesh panels on metal frames. The white fixture to the left is the mount for the coaxial machine gun; to the front of the mount, daylight can be seen through the port in the mantlet for the machine gun. *Chris Hughes*

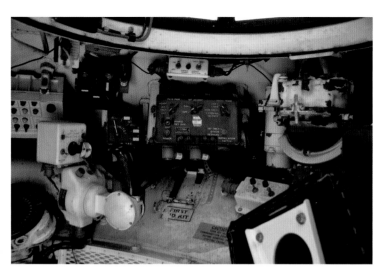

The commander's position looking to the right. The mechanism with the lever at lower left is the commander's control handle, while the box above it is the master control for the grenade launcher system. The box at center is the AM-1780/VRC audio frequency amplifier—the master control for the radio and intercom systems. His seat is folded to the lower right. *Chris Hughes*

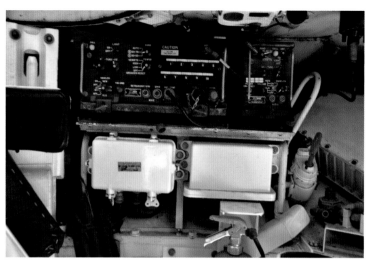

Located in the rear of the turret is the M551's radio system. It is made up of two separate assemblies: the RT-246/VRC transmitter on the left and the R-442/VRC receiver on the right. Below the radio are the amplifier integrator (*left*) and the power supply (*right*). *Chris Hughes*

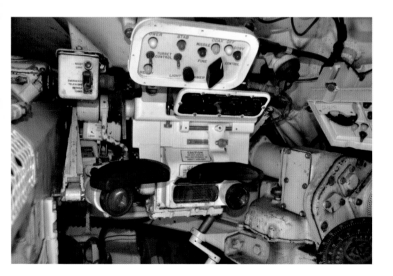

The gunner's controls. Two sighting devices dominate the array: on the left is the M127 telescope and on the tight is the XM44 periscope, with its control mechanism being the rectangular black panel. At the top is the gun and turret control selector panel. *Chris Hughes*

A horizontal ammunition rack located beneath the gun launcher holds five 152 mm rounds. To its right (left in the photo) is one of two air cylinders for the CBSS system. *Chris Hughes*

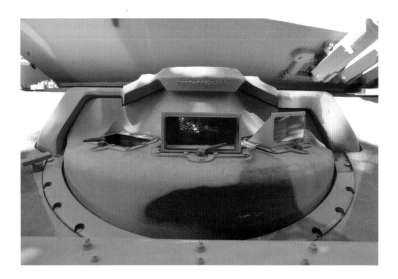

The driver's rotating hatch in the closed position. It is located in the front center of the hull. When closed, the driver used three M47 periscopes to view his path. Three small wipers kept the periscopes clean, and washing fluid could be sprayed on them from nozzles near the wiper arm base. *Chris Hughes*

The view looking down into the driver's hatch shows the conventional ammunition stowage rack to his left. Missiles were intended to be stowed in the rack to his right. Boxes of 7.62 mm ammunition for the coaxial machine gun sit in brackets directly behind his seat. *Chris Hughes*

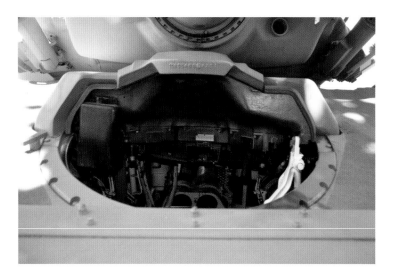

One of the unique aspects of the M551 was that to open, the driver's hatch rotated around the driver's head. It was attached from a central pivot point and was aligned on a circular track. The dark boxlike object to the left is the reservoir for the periscope washer fluid. *Chris Hughes*.

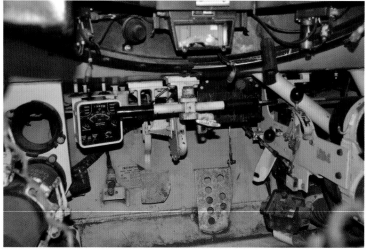

Looking forward from the driver's seat are his controls. The T-shaped object in the center is the steering bar, and the brake is below on the left, with the accelerator pedal on the right. The headlight dimmer switch is seen to the left of the brake, much like on cars of the 1960s. The black vertical bar is the hatch cover handle grip, and the transmission shift lever is below that and to the right. *Chris Hughes*

This is one of two compensating idler wheels located on the front of the M551. Each was paired with a hydraulic track adjuster. Adding or removing grease to the adjuster allowed it to vary the track tension. *Chris Hughes*

The drive sprockets for the M551 are located on the rear of the tank. They are unique in that the sprocket itself is sandwiched between two carrier sprocket wheels, which keep the sprocket aligned on the tracks. Visible to the right of the last road wheel is one of fourteen brackets (seven on each side) used to secure the tank to an airdrop platform. *Chris Hughes*

A total of twenty rubber-tired road wheels were mounted on the M551. They were arranged in sets of two, with five on each side, and were connected to internal torsion arms. A grease fitting was installed in the center of each wheel and was normally painted red, but it has been obscured here by grease. *Chris Hughes*

The M551 was shod with T138 single-pin tracks. Each track link contained two rubber pads on the face. They are nearly worn off on the MVTF examples, causing the track faces to show pronounced wear. The bolted-on idler housing is seen to the left. A flexible rubber mudguard keeps debris down on the upper hull. *Chris Hughes*

CHAPTER 3
Field Use

Specifications	
Model	M551
Weight*	33,460 lbs.
Length**	248.3 inches
Width**	110 inches
Height**	150 inches
Crew	4
Maximum speed	45 mph
Fuel capacity	158 gallons
Range	350 miles
Electrical	24-volt negative ground
Transmission speeds	4 forward 2 reverse
Turning radius feet	Pivot
Armament	
Main	1 x 152 mm
Secondary	1 x .30 cal.
Flexible	1 x .50 cal.

* Fighting weight
** Overall dimensions, measured with main gun facing forward and antiaircraft machine gun mounted

Engine Data	
Engine make/model	General Motors 6V53T
Number of cylinders	6
Cubic-inch displacement	318.6
Horsepower	300 @ 2,800 rpm
Torque	615 @ 2,100 rpm

While the Sheridan saw its greatest combat use in Vietnam, it was also deployed to Panama in Operation Just Cause and finally in Desert Storm. It was during the latter that the Shillelagh was finally fired in combat, although only a half dozen of the 85,000 missiles produced were used, and those were against bunkers.

This was in marked contrast to the thousands of conventional rounds expended in Vietnam, Panama, and Desert Storm. However, the massive bore of the 152 mm gun when combined with its short barrel meant that when firing a conventional round, the first one or two road wheel positions would lift from the ground.

Even though no suitable air-transportable replacement for the Sheridan had been created, in 1978 the decision was made to withdraw the Sheridan from combat units, except for the 82nd Airborne Division (which ultimately deployed to Panama and Desert Storm with the vehicles). The remaining M551 vehicles retained in Army inventory were transferred to the Desert Training Center, where the vehicles were visually modified to resemble Soviet vehicles for use by opposing forces during training scenarios. In 2003, the vehicles were retired from even this use.

An armored shield with folding top panels is visible on M551 registration number 12D67468 from Troop B, 3rd Squadron, 4th Cavalry, 25th Infantry Division, at Cu Chi, Republic of Vietnam, on January 27, 1969. The support bars originally mounted under the smoke-grenade projectors had long since been removed from most M551s. *National Archives*

This row of M551s at Cu Chi on January 24, 1969, was part of a group of fifty-four Sheridans that had recently arrived in Vietnam and had yet to be fully equipped and readied for combat operations. They were assigned to the 3rd Squadron, 4th Cavalry. The closest vehicle is registration number 12C79568. *National Archives*

Several members of Troop B, 3rd Squadron, 4th Cavalry, are preparing a newly arrived M551 for combat operations at Cu Chi on January 27, 1969. Two of them are securing a brand-new shovel to the rear of the hull. Note the webbing straps for securing the covers of the flotation barrier, and the white stencils indicating the locations for stored items. *National Archives*

In concert with several M113 armored personnel carriers, M551 Sheridan registration number 12C67868, from Troop A, 11th Armored Cavalry Regiment, is patrolling the outskirts of Long Binh around January 1969. Attached to the rear of the turret is a ramshackle rack full of containers, rations boxes, and other gear, and the cupola is equipped with an ACAV shield. *National Archives*

An M551 Sheridan with an ACAV shield and several M113 armored personnel carriers are moving out of the base of Troop A, 11th Armored Cavalry, on a sweep of the area around Long Binh circa January 1969. This appears to be the same Sheridan seen in the preceding photo, judging by the arrangement of stored equipment and the design of the rack on the turret. A xenon searchlight with a dustcover over it is mounted on the turret. *National Archives*

Another M551 serving with the 11th Armored Cavalry around January 1969 is making its way through the bush during a sweep near Long Binh. The improvised rack on the turret appears to be constructed of welded angle irons and contains two plastic liquid containers and other supplies. Several folding cots are stored on the vehicle. *National Archives*

During a sweep of the Long Binh area around January 1969, an M551 from Troop A, 11th Armored Cavalry Regiment, is advancing through the bush. Unlike some other Sheridans of this troop at that time, this vehicle was not equipped with an improvised rack for equipment on the rear of the turret. *National Archives*

An M551 Sheridan, registration number 12C76768, from the 3rd Squadron, 4th Cavalry, is proceeding from the motor park to Range Number 3, south of the perimeter at Cu Chi on February 1, 1969. Mounted on the front of the hull is a metal frame supporting chain-link fencing, designed to detonate enemy rocket-propelled grenade (RPG) antitank rounds before striking the light armor of the front of the hull.
National Archives

Troopers from the 3rd Squadron, 4th Cavalry, prepare to move their M551s into firing positions at Range Number 3 at Cu Chi on February 1, 1969. Chain-link RPG shields are on the fronts of the first two vehicles; their support frames are made of angle irons that have been bolted together. The second vehicle bears registration number 12C76768. All the gun launcher barrels have white bands. Far down the line is a gun launcher barrel with the nickname "HARD CORE 7" painted to the front of the white band. *National Archives*

Considerable dust has been churned up by the firing of the gun launchers of these M551s from Troop B, 3rd Squadron, 4th Cavalry, during firing practice at Range Number 3, Cu Chi, on February 1, 1969. It is clear from these photos that some but not all of the Sheridans of the 3rd Squadron were equipped with chain-link RPG guards on their front ends. *National Archives*

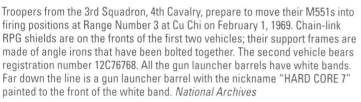

M551 Sheridan crews of the 3rd Squadron, 4th Cavalry, are engaged in live-fire practice during the unit's final day of training at Range Number 3 at Cu Chi on February 1, 1969. Red, green, and yellow flags flying from the turrets indicate the status of the individual vehicles' weapons and gun launcher crews. *National Archives*

The dust has settled in this image of the same lineup of M551 Sheridans seen in the preceding photo, viewed from the same perspective at Cu Chi on February 1, 1969. On the closest vehicle is registration number 12D67468. Most if not all of these Sheridans were equipped with armored shields with folding tops for the cupola-mounted .50-caliber machine guns. *National Archives*

In another view from the February 1, 1969, live-fire training exercise by the 3rd Squadron, 4th Cavalry, at Range Number 3, Cu Chi, the gun launcher of the M551 Sheridan at the center has just fired a round, causing the vehicle to rear back on its suspension. *National Archives*

M551 registration number 12C68368 from the 3rd Squadron, 4th Cavalry, makes its way through tall grass during live-fire exercises at Cu Chi on February 1, 1969. The helmet of the driver's combat vehicle crewman (CVC) is lying to the front of his hatch on the glacis. The nickname "SWEET X SUE" is painted in red on the aft part of the sponson. A small, red-over-white cavalry guidon is on the sponson between the fire-extinguisher handle and the vehicle's registration number. *National Archives*

This and the following four photos depict M551 registration number 12C67368, nicknamed "THE 'DEVIL AVENGES,'" from Troop B, 11th Armored Cavalry Regiment, in February 1969. The nickname was painted on the sponson. Note the ACAV shield, the three rings around the barrel of the gun launcher, and the equipment rack made of different gauges of angle irons on the rear of the turret. *National Archives*

The left access panel on the engine deck, actually two panels with a hinge between them, is open in this left-rear view of "THE 'DEVIL AVENGES.'" Several folding cots are stored on the rear of the hull. *Patton Museum*

In a frontal view of "THE 'DEVIL AVENGES,'" two rations boxes are secured to the glacis next to the headlights, and a thermal hot/cold food container is stowed on the side of the turret. Toward the bottom of the frontal plate of the lower hull, the curved front end of an armor plate added to give the bottom of the vehicle extra protection is visible. There are cutouts on the top edge of the plate for the hex bolts that secure the front of the plate. *Patton Museum*

Several crewmen of "THE 'DEVIL AVENGES'" are enjoying the scant shade afforded by the vehicle. They also have rigged a shelter to the forward part of the vehicle to provide all the relief they can get from the intense Southeast Asia sun. *Patton Museum*

"THE 'DEVIL AVENGES'" is observed from the left front, showing the rigged shelter on the right side, the ACAV shield, the front of the add-on armor kit for the belly, and other details. In the left background are several M113 armored personnel carriers with ACAV shields. *Patton Museum*

An M551 Sheridan nicknamed "HARD CORE 7," seen in an earlier photograph, from Troop D, 3rd Squadron, 4th Cavalry, is moving into position in the bush outside Cu Chi on February 12, 1969. The vehicle is equipped with chain-link armor to the front of the hull; the fencing material is in a battered state. Note the rusty condition of the angle-iron frame of the chain-link armor. The machine gun on the cupola is equipped with the early-type armor with the folding panels at the top. *National Archives*

"HARD CORE 7" is seen from a different perspective as it advances to a firing position outside Cu Chi on February 12, 1969. The number "27" is painted on the sponson in large numerals. The yellow marking on the rear part of the gun launcher barrel appears to be a likeness of a knight chess piece. *National Archives*

Two M551s assigned to Troop D, 3rd Squadron, 4th Cavalry, have opened fire on a Vietcong position in the vicinity of Cu Chi on February 12, 1969. Stored on the rear of the hull of the closer Sheridan are rolls of barbed wire and chain-link fence, for perimeter defense at night. A fire, typical of the types started by tracer rounds, has begun to smolder near one of the vehicles. *National Archives*

The same two M551s shown in the preceding photo are seen from a closer perspective at the same location and date. A yellow knight chess piece is faintly visible, painted on the gun launcher barrel. *National Archives*

Two M551 Sheridans from Troop C, 3rd Squadron, 4th Cavalry, cross a field during a search-and-destroy mission some 8.7 miles northwest of Cu Chi on February 22, 1969. The vehicle on the left bears the number "27" on its sponson and has an improvised equipment rack on the rear of the turret, including a holder for a five-gallon liquid container. *National Archives*

The same two M551s depicted in the preceding photograph, from Troop C, 3rd Squadron, 4th Cavalry, have moved closer to the photographer during a search-and-destroy mission out of Cu Chi on February 22, 1969. *National Archives*

An M551 Sheridan is supporting troops from the 1st Squadron, 11th Armored Cavalry, during a search for hidden Vietcong fighters near Long Binh Post, Republic of Vietnam, on February 23, 1969. Two plastic liquid containers are stashed on the rack on the rear of the turret. *National Archives*

An M551 Sheridan from Troop C, 3rd Squadron, 4th Cavalry, is flying an American flag from a whip antenna while accompanying an M113 armored personnel carrier across an embankment near Boi Loi Woods, Republic of Vietnam, in February 1969. The M551 has the .50-caliber gun shield with the folding top panels; the number "25" is on the sponson. *Patton Museum*

The 152 mm gun launcher of M551 Sheridan registration number 12C77068 has just fired a round at a suspected Vietcong position on February 22, 1969. The vehicle was attached to Troop C, 3rd Squadron, 4th Cavalry, 25th Infantry Division. The small, flap-shaped object above the "U.S. ARMY" inscription is the left flotation-barrier step, a folding step to prevent damage from tromping on the flotation-barrier cover. A similar step was on the right sponson. *National Archives*

During the Operation Acid Test I, Punch Card V winter war games in Alaska in February 1969, a snow-covered M551 with Company D, 40th Armored Regiment, 172nd Infantry Brigade, moves out to engage aggressor forces. Among other markings on the sponson are the registration number, 13E933, and a yellow number "21." *National Archives*

A Light Recovery Vehicle M578 is beginning to tow a disabled M551, registration number 13E927, from Company D, 40th Armored Regiment, during Operation Acid Test I, Punch Card V, in Alaska on February 11 or 12, 1969. The Sheridan had stalled in the cold weather and couldn't be restarted. *National Archives*

Crewmen are preparing an M551 Sheridan, registration number 12C81868, for its next mission at a base in Vietnam. In addition to the usual .50-caliber and 7.62 mm ammunition boxes strapped to the turret, several of what appear to be 40 mm ammunition boxes are secured to the glacis and to the turret above the smoke-grenade projectors. These boxes would have been useful for storing miscellaneous gear, protecting it from dampness and dust. *Chris Harlow*

Using the winch and boom of a Medium Recovery Vehicle M88, technicians are removing the turret from an M551 Sheridan in Vietnam during June 1969. The white parts of the turret basket are visible just above the top of the hull. *Rock Island Arsenal Museum*

Once the turret has been removed from an M551 and the vehicle has been driven out of the way, the Medium Recovery Vehicle M88 is lowering the turret assembly to the ground in front of a shelter. On the rear of the M88's superstructure is secured a Mermite can—a thermal hot-and-cold food container. *Rock Island Arsenal Museum*

The turret assembly seen in the two preceding photographs has been placed on a special stand, where technicians will be able to work on it. The feature with the five holes on the front of the floor of the turret is part of a rack for conventional 152 mm ammunition. *Rock Island Arsenal Museum*

An M551 Sheridan is viewed from above during testing at Aberdeen Proving Ground, Maryland, on July 30, 1969. This vehicle was equipped with the early type of armored shield for the commander, with foldable upper panels. The object on the turret roof to the front of the ventilator hood is the receptacle for the power cable from the searchlight, when installed. *National Archives*

This M551, likely the same one shown in the preceding photo, is being subjected to cross-country tests at Aberdeen Proving Ground on July 30, 1969. The early-type shield for the cupola machine gun provided some ballistic protection from the front, but there were gaps in the protection, such as below the gun mount.
National Archives

An M551 Sheridan from Troop A, 3rd Squadron, 4th Cavalry, 25th Infantry Division, advances across solid ground in a rice paddy during a search-and-destroy mission in the Republic of Vietnam on July 8, 1969. Note the clutter, including a plastic liquid container and several rations boxes, on the turret roof to the front of the cupola, and the improvised storage rack holding ammunition boxes on the rear of the turret. Registration number 12C84268, a small red-and-white cavalry guidon, and the number "16" are on the sponson. *National Archives*

Troopers of the 3rd Platoon, Troop C, 3rd Squadron, 4th Cavalry, are preparing an overnight laager site for night defensive perimeter (NDP) for their M551 Sheridan near the Hobo Woods in the Republic of Vietnam on July 15, 1969. A section of chain-link fence with a post driven into the ground to support it is visible to the left front of the vehicle; this would form part of a barrier that would detonate any incoming RPG rounds before they could strike the tank. As a field modification, a storage rack has been installed above the rear of the hull. *National Archives*

Concertina wire is in the foreground along the perimeter of Landing Zone Hampton, as an M551 Sheridan with a yellow number "25" on the sponson, followed by an M113, moves through the muddy ground on August 3, 1969. The Sheridan was assigned to Troop A, 3rd Squadron, 4th Cavalry. Note the yellow band painted around the gun launcher barrel. *National Archives*

Members of the crew of M551 registration number 12C77869 are relaxing on the vehicle, which is sitting in a mudhole at Landing Zone Hampton. The vehicle was with Troop A, 3rd Squadron, 4th Cavalry, and the photo was taken on August 3, 1969. The vehicle's nickname, "SUDDEN DEATH," is painted in white on the gun launcher barrel; a red band also is on the barrel. Note the locally made storage rack on the rear of the turret, welded from reinforcing rods and with chain-link fencing material on the rear and bottom. *National Archives*

Specialist 4th Class Nikola Andrejenko, a track mechanic, is at the driver's controls of an M551 Sheridan at the US Army Supply Depot at Long Binh, Republic of Vietnam, in early August 1969. The registration number is 12C90568. Of interest is the neatly built storage rack on the turret: although one end of the rack is at the right-rear corner of the turret, on the far side of the turret the top of the rack is visible, terminating approximately at the center of that side of the turret.
National Archives

Personnel from the 185th Maintenance Battalion are servicing two M551 Sheridans with ACAV shields on the turrets at the maintenance yard at Quan Loi, Republic of Vietnam, on December 22, 1969. The Sheridan on the left, registration number 12C98468, has a faded nickname, "ARETHA," painted on the sponson.
National Archives

Several members of Troop E, 2nd Squadron, 11th Armored Cavalry Regiment, are checking their mine detector before a mission, at Fire Base Eunice, Republic of Vietnam, on December 25, 1969. Next to them is an M551 Sheridan; this is one of the vehicles equipped with a locally constructed equipment bin, the left side of which terminated at about the center of the left side of the turret. *National Archives*

More of the same M551 shown in the preceding photo is visible in this companion view. A nickname, which unfortunately is illegible, is faintly visible on the sponson, painted in large letters. An M60 7.62 mm machine gun is propped up atop the turret. *National Archives*

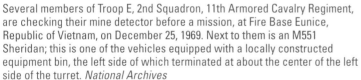

A minesweeping team from Troop E, 2d Squadron, 11th Armored Cavalry, is operating in front of two M551 Sheridans and several M113s during a mission out of Fire Base Eunice on December 26, 1969. Both Sheridans have the ACAV shields. *National Archives*

In the Quan Loi area of the Republic of Vietnam, an M551 Sheridan with ACAV shield is part of a force conducting a search-and-destroy mission on December 28, 1969. This vehicle has the locally fabricated wraparound equipment bin on the turret. In the background, several M113 armored personnel carriers are operating. *National Archives*

In a photo taken during the same mission and date as in the preceding photo, crewmen of an M113, *left*, and an M551 are relaxing during a break in the action. The original caption of the photo identifies the unit only as "Troop B, 2nd Squadron." A large number of cartons of unknown contents are stacked on the rear of the hull of the M551. *National Archives*

The commander and the gunner of an M551 from Troop C, 1st Squadron, 11th Armored Cavalry, are on the alert as their vehicle advances through the bush outside Fire Support Base Dennis, 3.7 miles north of Loc Ninh, Republic of Vietnam, on January 22, 1970. This Sheridan is furnished with an ACAV shield and a locally fabricated equipment bin on the rear and left side of the turret. This bin was made of expanded-steel mesh on a metal frame and was strong enough to support not only a sizable amount of gear but also four plastic liquid containers. *National Archives*

During the Cambodian incursion in May 1970, two M551s from the 11th Armored Cavalry Regiment are escorting a pair of M113s on a road near Snoul. *National Archives*

Two troopers from the 11th Armored Cavalry Regiment are surveying the damage to an M551 Sheridan, suffered in a mine explosion on Highway 7 outside Snoul, Cambodia, on May 8, 1970. The blast caused the track to come off and resulted in severe buckling to the sponson. *National Archives*

Another photo of the M551 that struck a mine on Highway 7 near Snoul, Cambodia, on May 8, 1970, shows more of the extent of the damage to the left side of the vehicle. In the foreground is a wheel assembly that was blown off the chassis. *National Archives*

Extra personnel are hitching a ride on an M551 Sheridan from Troop A, 3rd Squadron, 4th Cavalry, during a sweep through the Cambodian jungle on May 18, 1970. Two spare bogie wheels are lying on the glacis to either side of the driver's hatch. *National Archives*

Taken on the same date and general location as the preceding photo, wary troops of 3rd Squadron, 4th Cavalry, ride on the rear of a Sheridan through thick jungle in Cambodia. Piled on the engine deck are plastic liquid containers, wooden crates filled with oddments, a 40 mm ammunition box, a roll of matting, and a tow bar. *National Archives*

Two M551 Sheridans are parked in a revetment at the Class 7 Area of the 625th Supply and Service Company in June 1970. The nearest vehicle is registration number 12F39169. Note the lack of machine gun mounts on the cupolas. What appears to be a turret stowage rack is lying on the engine deck of the first M551 and on the glacis of the second vehicle. *National Archives*

An M551 Sheridan from Troop C, 3rd Squadron, 4th Cavalry Regiment, is passing through a security checkpoint while departing from Cambodia back to the Republic of Vietnam in June 1970. A large amount of gear, including a number of 20 mm or 40 mm ammunition boxes (or both sizes), are secured to the rear of the vehicle. *National Archives*

While ascending a steep slope above a muddy stream, an M551 assigned to Troop B, 1st Squadron, 1st Cavalry, Americal Division, is attempting to tow another vehicle that got mired in the stream, east of Landing Zone Hawk Hill, on August 15, 1970. The M551 is viewed from the rear; note the roll of chain-link fence on the bow, for setting up anti-RPG defenses at night. *National Archives*

In a photo related to the preceding one, an M551 from Troop B, 1st Squadron, 1st Cavalry, is climbing up a steep riverbank, east of Landing Zone Hawk Hill, on August 15, 1970. An M113 is operating in the background. *National Archives*

An M551 serving with Troop B, 1st Squadron, 1st Cavalry, Americal Division, has taken up a firing position east of Landing Zone Hawk Hill on August 15, 1970. This vehicle had one of the equipment racks on the turret that was made of expanded-steel mesh on a metal frame, with the left side of the rack ending at about the center of the left side of the turret. *National Archives*

To the right, a trooper from the 1st Squadron, 1st Cavalry, is leading an M551 Sheridan across a rice paddy in the vicinity of Landing Zone Hawk Hill on August 15, 1970. This vehicle has a baggage rack on the rear and sides of the turret, and an extra, improvised rack above the smoke projectors, and has a ACAV shield mounted upside down as protection for the TC. *National Archives*

An M551 from the US 1st Infantry Division (Mechanized) is passing the reviewing stand during a demonstration at the start of Phase II-FTX "Certain Thrust," part of the annual Exercise REFORGER (Return of Forces to Germany), in early October 1970. This vehicle has the early-type shield for the machine gun on the cupola, and a two-color camouflage has been applied. *National Archives*

Following hand signals from a soldier in an arctic parka, the crew of an M551 from Troop E, 16th Cavalry, 171st Infantry Brigade, is preparing to move out on a routine road patrol during Operation Acid Test Three, a cold-weather readiness exercise in Alaska, on December 6, 1970. The tow shackle on the right side of the bow has been painted yellow. *National Archives*

An M551 severely damaged by a mine explosion lies in an equipment yard in Vietnam. This vehicle was equipped with supplemental armor under the front of the hull, the front end of which is visible at the bottom of the bow, and, although part of the suspension was destroyed, the extra armor helped preserve the crew. *William Powis*

This M551 exhibits a style of locally fabricated storage rack on the rear of the hull that was encountered on Sheridans of the 2nd Squadron, 11th Armored Cavalry Regiment. Also, a locally made, wraparound storage rack is on the turret. *William Powis*

The same Sheridan is viewed from the left front. This vehicle was from 1969 production; subsequently, it was fully repaired and continued to serve until it was retired from service in 1985. *William Powis*

In Vietnam around 1970, an M551 Sheridan has been loaded onto an M15A2 semitrailer coupled to an M123 truck-tractor, for transport. The turret is traversed to the rear, and a cover has been placed over the muzzle of the gun launcher. *Chris Harlow*

A Medium Recovery Vehicle M88 is parked in a muddy, fortified base camp next to an M551 Sheridan equipped with an ACAV shield and a locally fabricated storage bin on the turret. *William Powis*

Around 1970, an M551 Sheridan is crossing an engineer bridge in the Republic of Vietnam. Although the photo is grainy and somewhat out of focus, it is clear that in addition to the commander's .50-caliber machine gun, the gunner is manning a .50-caliber machine gun mounted on the left side of the turret roof, and that gun is pointing slightly downward and to the vehicle's left. *William Powis*

Two M551 Sheridans from Troop A, 2nd Squadron, 4th Cavalry, 4th Armored Division, are viewed through a telephoto lens as they advance to a firing range at the Seventh Army Training Center at Grafenwoehr, Germany, in or around 1970. *Patton Museum*

The commander of an M551, registration number 12B38770, from the 11th Armored Cavalry Regiment is holding up a captured AK-47 during a sweep in or around 1970. The nickname "Marcia" is painted in white on the ACAV shield, and "Birth-Control" is painted in white on the gun launcher barrel. The rear and side components of the ACAV shield often were used for attaching supply crates and boxes.
Dave Orth

A Light Recovery Vehicle M578 is dragging an M551 that has lost its left running gear to a land mine, in Vietnam around 1970. The bogie wheels that were blown off in the blast are piled on the glacis and on the turret. The vehicle's troop letter and order-of-march number, G39, is stenciled in white on the door of the optical transmitter.

Faintly visible on the sponson of M551 registration number 12F37369, from the 11th Cavalry Regiment, is the nickname "BLESSED ONE." The gunner has been provided with an M60 machine gun with an ACAV shield, while the commander stands behind a .50-caliber machine gun with the larger-sized ACAV shield. The photo was taken near Hung Loc, Republic of Vietnam, on January 23, 1971. *National Archives*

The same M551 from the 11th Armored Cavalry Regiment is viewed from the front, with the turret traversed to the front. Supplementary armor is visible on the lower part of the bow. Very faintly visible on the gunner's machine gun shield is graffiti, including "TO RITA WITH LOVE." *National Archives*

"THE REBEL" is the nickname painted on the 152 mm gun launcher of M551 registration number 12C72168, outside Tam Ky, Vietnam, in March 1971. The gun shield is the early type with folding flaps on top. A large storage box has been mounted on the side of the turret. *Patton Museum*

A mascot dog is seated on the turret of an M551 that has been modified with two .50-caliber machine guns mounted on the forward part of the turret roof. In the combat environment of Vietnam, sometimes the focused destructive power of a pair of .50s was preferable and more economical than that of the 152 mm gun launcher. *National Archives*

A crewman is standing on the left brush guard of a Sheridan from Delta Company, 1st Battalion, 1st Training Brigade, during training at Fort Knox, Kentucky, in June 1971. A yellow placard with the troop letter and order-of-march number, "D-33," is on the sponson. *Doug Kibbey*

D-33 is viewed from the left side at Fort Knox in June 1971. Note the three yellow rings on the muzzle. *Doug Kibbey*

Another Sheridan from Delta Company, 1st Battalion, 1st Training Brigade, marked "D-44" on a yellow patch on the sponson, precedes an M113 armored personnel carrier during an exercise at Fort Knox in June 1971. *Doug Kibbey*

The entire left side of D-33, from Delta Company, 1st Battalion, 1st Training Brigade, is displayed at Fort Knox in June 1971. *Doug Kibbey*

An M551 Sheridan, registration number 12C80768, from Troop E, 17th Cavalry, 173rd Airborne Brigade, has paused in a palm grove in Vietnam in August 1971. Note the five-gallon metal liquid container above the smoke projectors on the right side of the turret, the rations boxes strapped above the left-hand projectors, and the section of four spare track links on the left side of the glacis. *Patton Museum*

An M551 Sheridan is parked on a dock at the Da Nang deepwater port in the Republic of Vietnam on November 8, 1971. Next to the Sheridan is a Cargo Carrier M548 with the cab top and windshield removed. In the background, a 6 x 6 wrecker truck is being hoisted aboard the freighter *John B. Waterman. National Archives*

In another view of the *John B. Waterman* at Da Nang on November 8, 1971, an M551 Sheridan, *left*, and two M48 tanks are awaiting loading on the ship. Lying on the dock to the front of the M551 is an ACAV shield for the commander's .50-caliber machine gun and other components. The covers for the flotation barrier are open and have been damaged, the tracks are missing, and the trim vane has been removed. *National Archives*

Advancing through woodlands during a training excercise at Fort Stewart, Georgia, in January 1972 are two M551s from the 82nd Airborne Division. They have the early-style ballistic shields for the commander, along with armored shields to the rear of the cupola. *Patton Museum*

An M551 Sheridan with markings for Troop G, 11th Armored Cavalry Regiment, is exiting a fuel and supply point in the Tan Uyen District during February 1972. Note the A-frame-style machine gun support, like the one used on the commander's cupola, on the left front of the turret roof, on which is mounted a .50-caliber machine gun for the gunner's use. *Doug Kibbey*

The turret of the same M551 shown in the preceding photo, number "37" from Troop G, 11th Armored Cavalry, is viewed close-up, showing the gunner's .50-caliber machine gun from another angle. *William Powis*

Locally constructed racks are present on the rear of the turret and the rear of the hull on this M551 from Troop G, 11th Armored Cavalry Regiment. A roll of chain-link fence is also stored on the rear of the hull, for nighttime defense against RPGs. *Doug Kibbey*

A soldier is refueling an M551 with several interesting modifications. On the left side of the turret roof, an A-frame-type machine gun mount of the type used on the commander's cupola has been installed, and it is supporting an M60 7.62 mm machine gun, which is fed by ammunition from a 40 mm ammunition box. Fitted over the exhaust next to the soldier is an extension made from an artillery powder container. *Doug Kibbey*

Painted in a four-color camouflage scheme that preceded the MERDC camouflage by a couple of years, an M551 with registration number 13F108, from Troop B, 3rd Battalion, 12th Infantry, 3rd Armored Division, is at a firing range at Grafenwoehr, West Germany, on May 16, 1973. The crew were taking their annual gunnery qualification course. *National Archives*

Another Sheridan from Troop B, 3rd Battalion, 12th Infantry, is on a firing line at Grafenwoehr on May 16, 1973, during crew gunnery qualifications. Along with the four-color camouflage, the vehicle had the new, low-visibility black recognition stars. On the rear of the turret is a simple rack for baggage, secured in position with straps. *National Archives*

Sheridans from Troop B, 3rd Battalion, 12th Infantry, are lined up on the firing line at Range 5 during gunnery qualifications at Grafenwoehr on May 16, 1973. All vehicles have the early frontal shield for the commander and the rear shield as well.
National Archives

The firing line at Grafenwoehr with Sheridans at the ready on May 16, 1973, is depicted. The .50-caliber machine guns are dismounted from the cupolas.
National Archives

An Armored Command Post Carrier M577 precedes three M551 Sheridans en route to their initial positions at the beginning of Exercise REFORGER, in West Germany in September 1974. At least the first two Sheridans lack shields for the cupola-mounted machine guns. *National Archives*

A local man and his children watch as members of the 2nd Cavalry take a meal break on their M551 near Alensheim, West Germany, during Exercise REFORGER in September 1974. The tank is painted in four-color camouflage, and it has the early-style ballistic frontal shield for the commander along with a rear shield. *National Archives*

A newly rebuilt M551 Sheridan is being put through firing tests at Fort McClellan, Arkansas, on February 20, 1975. Conventional rounds for the 152 mm gun launcher are lined up on the tarpaulin toward the left. *National Archives*

Turrets traversed to the rear, an M551A1 and a Main Battle Tank M60 are returning from firing tests of their main weapons at Fort McClellan, Arkansas, on March 21, 1975. The conduit for the electrical lines from the laser rangefinder's transmitter-receiver, under the commander's gun mount, back to the electronics boxes inside the commander's rear shield, is noticeable. A good view is available of the rear shield; a door is built into the rear armor, with two hinges on the bottom and a latch bolt on each upper corner. *National Archives*

A Sheridan, registration number13E341, with Company A, 4th Battalion, 68th Armored Battalion, 82nd Airborne Division, is returning to the command post after an assault mission during a training exercise at Fort Bragg, North Carolina, on November 4, 1975. As may be seen in this photo, when the early-style shield for the commander was used, the open hatch doors provided him with protection from the sides, and the rear armored shield gave him protection from the rear and, partially, from the sides.
National Archives

The driver of an M551 Sheridan is visible in his open hatch at the Combat Equipment Group Europe facility in West Germany, where combat vehicles about to participate in Exercise REFORGER were activated, in late August 1977. Of note on the turret, which is turned to the rear, are the position of the equipment rack on the rear of the turret when not supported by its straps, and the open door on the commander's rear armored shield.
National Archives

An M551 Sheridan, unit and location unknown, displays a MERDC desert camouflage scheme in an August 1977 photograph. Part of the frontal shield for the commander is present, with the top flaps folded down, and, with the rest of the frontal shield and the machine gun not mounted, a bit of the inner surfaces of the rear shield are visible. *National Archives*

The same M551 shown in the preceding photograph is viewed from a lower perspective to the left front. Above the turret is the commander's .50-caliber machine gun cradle and ammunition-box holder.
National Archives

Clad in a parka, the commander of an M551 from the 11th Armored Cavalry is leaning on his .50-caliber machine gun while on guard duty along the border between East and West Germany in May 1979. This vehicle, registration number 12C67768, is also the subject of the next several photos. It has the early-style armored shield over the cupola. *Department of Defense*

Troopers of the 11th Armored Cavalry are deployed around the M551 Sheridan while guarding the same area of the border depicted in the preceding photo, in May 1979. Note how the four-color camouflage runs into the open hatch of the driver. *Department of Defense*

More of the four-color camouflage scheme of the M551 is visible in this posed view, wherein troopers are issuing out of an Armored Personnel Carrier M113 near a guard tower along the border between East and West Germany in May 1979. *Department of Defense*

Troopers wielding M16 rifles pose alongside M551 registration number 12C67768 at the border between the two Germanys in May 1979: a risky proposition, since advancing beyond the guardrail could result in death. *Department of Defense*

Two US soldiers portraying Soviet troops, including one on the ground poised with an RPG launcher, are posing with an M551NTC: a version of the Sheridan visually modified (VisMod) to resemble, from a distance, a Soviet armored vehicle, in this case a BMP infantry fighting vehicle. The M551NTCs were employed at the National Training Center, Fort Irwin, California, and at the Joint Readiness Training Center, Fort Polk, Louisiana. The occasion depicted in this photo was Exercise Gallant Eagle '82 at the National Training Center at Fort Irwin, California. In addition to a fake cannon barrel and antitank missile, the vehicle is equipped with MILES (Multiple Integrated Laser Engagement System) equipment for simulating and registering the effects of gunfire. *Department of Defense*

An M551A1 Sheridan of 3rd Battalion, 73rd Cavalry Regiment, 82nd Airborne Division, has just disembarked from Army utility landing craft 1674 (LCU-1674), at Vieques Island, Puerto Rico, during Operation OCEAN VENTURE '84. The transmitter/receiver of the M551A1's laser rangefinder is clearly visible below the commander's machine gun mount. In keeping with the 73rd Cavalry's nickname, "Airborne Thunder," the dustcover for the searchlight has a stencil of an M551 Sheridan silhouette with a thunderbolt. *Department of Defense*

Another Sheridan assigned to the 73rd Cavalry Regiment, 82nd Airborne Division, is being off-loaded from a C-130 Hercules cargo aircraft during Exercise Ocean Venture '84 at Vieques Island. A woodland MERDC camouflage has been applied to the vehicle. *Department of Defense*

The crew of an M551A1, operating as opposing forces, take five during an exercise at Fort Polk, Louisiana, in October 1985. The vehicle has MILES equipment, including a blank-firing attachment (BFA) fitted over the barrel of the commander's .50-caliber machine gun; a Hoffman Device gunfire simulator on the barrel of the 152 mm gun launcher; and, on the right front of the turret roof, an amber-colored strobe light inside a guard, for signaling simulated near hits and hits on the vehicle. *Department of Defense*

This M551NTC has been altered to resemble a Soviet tank of the mid-1980s, for use by the 177th Armored Brigade while operating as opposing forces during field exercises at the National Training Center, Fort Irwin, California, in October 1985. The sides of the turret were built out, and modifications were made to the front and the rear of the hull. The 177th Armored Brigade operated a number of M551s modified to simulate a variety of then-current Soviet armored vehicles. *Department of Defense*

An M551NTC Sheridan with MILES equipment and another visually modified armored vehicle in the background are negotiating desert terrain during an Opposing Forces exercise at the National Training Center in January 1986. *Department of Defense*

A column of M551NTCs crosses the desert during an Opposing Forces exercise at the National Training Center in January 1986. They are modified to resemble Soviet armored vehicles of different types. *Department of Defense*

An M551NTC from the 177th Armored Brigade, revamped to look like a Soviet BMP-1 infantry combat vehicle, is in position before a simulated battle at the National Training Center in February 1987. On the turret are simulations of the BMP-1's 73 mm gun and AT-3A Sagger A antitank wire-guided missile (ATGM). *Department of Defense*

Behind the M998 HMMWV in the foreground are two M551NTCs from the 177th Armored Brigade during a training exercise at the National Training Center in March 1988. The two M551s were altered to resemble then-current Soviet main battle tanks. *Department of Defense*

An M551NTC modified to resemble a Soviet ZSU-23-4 Shilka self-propelled antiaircraft gun system moves along a dirt road during an exercise at the National Training Center in March 1988. On the roof of the turret is a simulated radar antenna, while on the front of the turret are facsimiles of the original vehicle's four 23 mm 2A7 automatic cannons. *Department of Defense*

A desert camouflage paint scheme has been applied to this M551NTC in the guise of a Soviet ZSU-23-4 Shilka at the National Training Center in March 1988. The extended fronts of the fenders of the VisMod vehicles were cut out on the inner and outer sides to provide side visibility to the driver. *Department of Defense*

At the National Training Center during March 1988, an M551NTC disguised as a Soviet T-72 main battle tank exhibits MILES equipment, including a hit indicator strobe light on the turret roof, a Hoffman Device on the main gun barrel, and a BFA on the barrel of the commander's .50-caliber machine gun. *Department of Defense*

Troops from the 177th Armored Brigade advance through the desert at the National Training Center in an M551NTC modified to resemble a Soviet T-72 during March 1988. *Department of Defense*

Another M55NTC, altered to resemble a Soviet T-55 main battle tank, is in use by the 177th Armored Brigade at the National Training Center in March 1988. Despite the Soviet-style camouflage and markings, US unit markings are present on the bow for 1st Battalion, 73rd Armor, an element of the 177th Armored Brigade. *Department of Defense*

The same M551NTC simulating a T-55 main battle tank that is seen in the preceding photo is observed from the right front in dim light. At the National Training Center, the 177th Armored Brigade portrayed an imaginary Soviet unit, the Guards 60th Motorized Rifle Division, which was based on Red Army organization and doctrine. *Department of Defense*

A pair of M551NTC Sheridans from the 177th Armored Brigade, mimicking Soviet BMP-1 mechanized infantry combat vehicles, are advancing along a dusty trail during an exercise at the National Training Center in March 1988. *Department of Defense*

M551NTCs altered to resemble Soviet T-72 tanks are maneuvering through a sand- and gravel-filled ravine during an exercise at the National Training Center in March 1988. *Department of Defense*

After being off-loaded from a tank transporter, an M551A1 Sheridan of 3rd Battalion, 73rd Armor, 82nd Airborne Division, moves out as part of Exercise Task Force Dragon / Golden Pheasant, a show of force in March 1988 to discourage Nicaraguan forces from entering Honduras. The nickname "BIG DUKE 2" is painted in black on the gun launcher barrel. *Department of Defense*

A Sheridan and an HMMWV guard an intersection in Panama during Operation Just Cause in December 1988. In this operation, the first in which Sheridans engaged in combat since the Vietnam War, the dictator of Panama, Manuel Noriega, was overthrown. *Department of Defense*

When the United States and allied countries built up their armed forces in Saudi Arabia in Operation Desert Shield, the buildup phase of the campaign to drive the Iraqis out of Kuwait, which began in August 1990, some of the first US military vehicles to arrive there were M551 Sheridans, such as this one, which had just arrived in the C-5 Galaxy in the background. *Department of Defense*

In a lineup of Sheridans from Company A, 3rd Battalion, 73rd Airborne Armor Regiment, 82nd Airborne Division, the first vehicle is an M551A1 (TTS), a basic M551A1 in which an AN/VSG-2B tank thermal sight has been installed. On the ground next to the vehicle are conventional and Shillelagh munitions, liquid containers, and other basic-issue items. *Department of Defense*

Green knapsacks contrast with recently applied sand-colored paint on an M551A1 Sheridan from the 82nd Armored Division, as it advances to a firing range during a live-fire exercise during Operation Desert Shield in September 1990. As a safety measure, a red flag has been attached to a radio antenna to signal the status of the gun launcher. *Department of Defense*

A US soldier is on standby with an M224 60 mm mortar as an M551 Sheridan with much equipment stowed on the turret patrols a section of desert during a readiness exercise during Operation Desert Shield. Folding cots for the crew are lashed to the side of the turret. *Department of Defense*

A Sheridan from Troop B, 73rd Cavalry, 82nd Airborne Division, is backing up the ramp into a C-130
Hercules transport plane around the time of the commencement of Operation Desert Storm in early 1991.
Note the two sections of spare track links stored under the turret-bustle baggage rack.
Department of Defense

Two M551NTC Sheridans, altered to resemble a Soviet BMP (*left*) and a Soviet main battle tank, are parked side by side at the National Training Center at Fort Irwin, California, in October 2002. *Department of Defense*

The same M551NTC posing as a BMP seen on the left side of the preceding photo is viewed from a closer perspective; another M551NTC, modified to resemble a Soviet ZSU-23-4 Shilka, is to the left. *Department of Defense*

Also photographed at the National Training Center in October 2002 was this M551NTC reworked to resemble a Soviet M-1974 (2S1) 122 mm self-propelled howitzer. Although the M551NTC Sheridans with visual modifications were far from exact replicas of the Soviet vehicles they were mimicking, they were close enough for training purposes. *Department of Defense*

On display at the National Training Center in October 2002 was this M551NTC portraying a Soviet ZSU-23-4 self-propelled (SP), 23 mm antiaircraft gun system. *Department of Defense*

Numerous M551NTC Sheridans serving in the roles of Soviet main battle tanks and tracked armored vehicles are lined up at the National Training Center, in the Mojave Desert in San Bernardino County, California, during October 2003. Nicknames are painted in black on the barrels, including "DOGPOUND KILLA" and "DOG POUND GANG." *Doug Kibbey*

By October 2003, the remaining M551NTCs at the National Training Center were being deprocessed, or phased out of service. In this photo, taken during that month, a line of M551NTCs are being stripped of their disguise components preparatory to being retired. Note the imitation gun barrel lying next to the turret on the closest vehicle. *Doug Kibbey*

An M551NTC is viewed from the upper front while being deprocessed at the National Training Center in October 2003. On the glacis and the turret are imitation reactive-armor tiles of the type used on Soviet main battle tanks. *Doug Kibbey*

The M551s were powered by the Detroit Diesel (GM) 6V53T, a V-6, two-cycle, supercharged engine with 318.4-cubic-inch displacement and maximum gross horsepower of 300 hp at 2,800 rpm. *Doug Kibbey*

Lying in the sand at the National Training Center in October 2003 is a nearly deprocessed M551NTC Sheridan chassis. Much of the outer shell of the hull has been stripped off, resulting in a rare view of the sloped inner shell of rolled 7039 aluminum-alloy armor. *Doug Kibbey*

Several M551NTCs, modified to resemble Soviet main battle tanks, are awaiting deprocessing at the National Training Center, Fort Irwin, California, in October 2003. Imitation reactive-armor tiles are visible on the glacis and turret of the closest vehicle. *Doug Kibbey*